Old MacDonald's FUN TIME FARM

Story by Gail Tuchman
Illustrations by Mary Ann Lloyd

HARCOURT BRACE & COMPANY

Orlando Atlanta Austin Boston San Francisco Chicago Dallas New York
Toronto London

2 Can you find this hen?

Can you find this mouse?

Can you find this cow?

4

Can you find this pig?

Can you find this sheep?

Can you find this horse?

hen

mouse

cow

pig

sheep

horse

Did you find them?

8